96 Leonard Street
London
EC2A 4RH

Franklin Watts Australia
14 Mars Road
Lane Cove
NSW 2066

ISBN 0 7496 2800 6

Dewey Decimal Classification Number 641.3

A CIP catalogue record for this book is available from the British Library

Editor: Samantha Armstrong
Designer: Kirstie Billingham
Consultant: Professor Paul Davies
Reading Consultant: Prue Goodwin, Reading and Language Information Centre, Reading

Printed in Hong Kong

What's for lunch?

Rice

What's for lunch?

Rice

Pam Robson

W
FRANKLIN WATTS
NEW YORK • LONDON • SYDNEY

Today we are having rice for lunch.
The rice we eat is usually white
but it can also be brown, red or black.
Rice is full of **starch**.
Eating rice gives you **energy**.

The rice plant is a kind of grass.
It grows best when it is mostly under water.
So rice is planted in fields
that are flooded with water.
These are called **paddy** fields.
Some rice is grown on hillsides
that are cut into **terraces**.

JOHN DEERE

Rice is grown in countries where it is hot
and there is plenty of rain.
In some places, farmers use machines
to help them grow their rice.
In other places, people do most of the work.

The plant grows
from **grains** of rice.
The grains grow into
green **seedlings.**
Farmers plant the seedlings
in the paddy fields.

Farmers water the rice plants
to help them grow.
They put **fertilizer**
on the fields
to make the plants strong.
Insects sometimes eat the rice.
The farmers use **insecticide**
to kill the insects.

Little flowers called **spikelets**
appear on the plant.
The grains of rice grow inside the spikelets.
Each grain is protected by a tough outside layer
called a **husk** and a skin of **bran.**

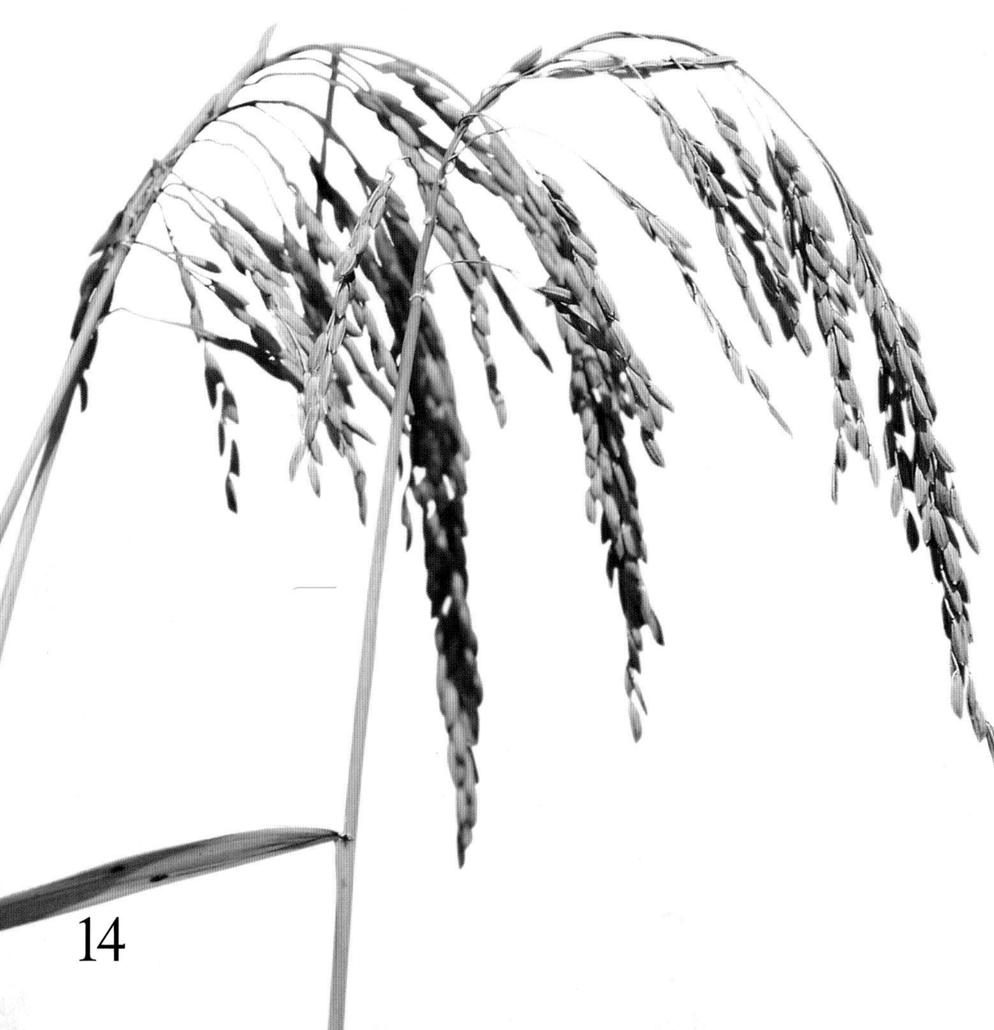

When the plants turn golden,
the rice is ready to **harvest**.
The farmers cut down the plants.
They separate the grains of rice from the plants.
This is called **threshing.**
Sometimes animals walk on the plants
to separate the grains of rice.

SUPREME

Next the farmers remove the husks
from the grains of rice.
A machine blows the husks away.
This is called **winnowing**.
Sometimes this is done in a factory.

At this stage the grains of rice
still have a layer of bran around them.
This is called brown rice.

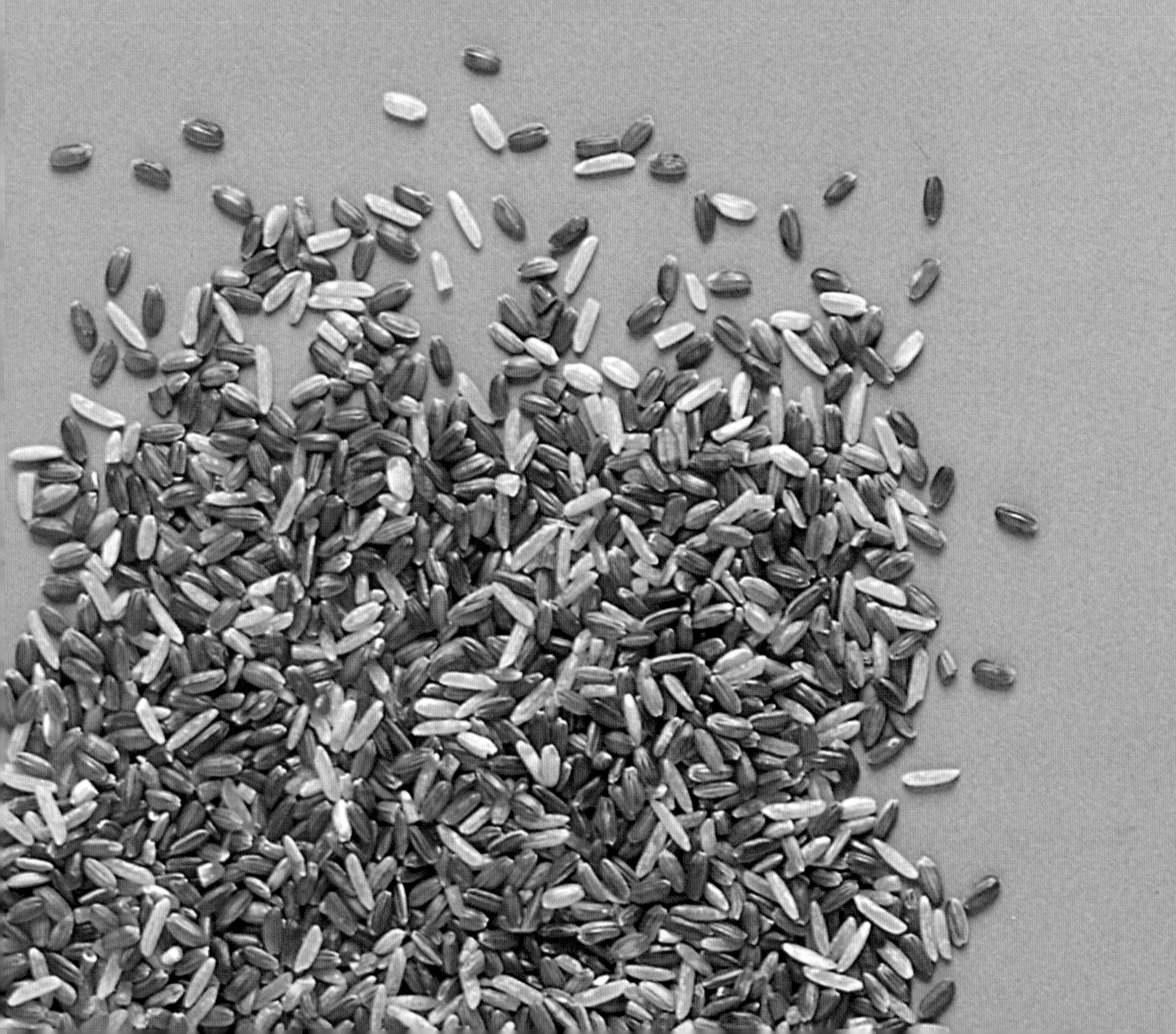

The bran can be removed.
This is called **polishing**.
Underneath the rice is white.

The rice is packed into sacks
and taken to shops and markets
where it is sold.
Some rice has a longer journey.
It is loaded onto ships and taken
to other countries to be sold.

S.W.L. 25 M.T.
PORT OF SACRAMENTO

There are many different types of rice.
Grains of rice can be long,
medium or short.
In India long grain rice
is eaten with curry.

In China long grain rice
is eaten with every meal.
Some types of rice have special names.
Basmati rice is very popular.

Rice can be cooked in many different ways.
It can be steamed and toasted
and used to make breakfast cereal.

People eat rice all over the world.
In China people use chopsticks
to eat their rice.
In India rice is sometimes
served on banana leaves.
For many people rice is
a very important food.

Glossary

bran	the brown skin of a grain of rice
energy	the strength to work and play
fertilizer	something that helps plants grow
grain	the small hard part of a grass that looks like a seed
husk	the outside covering of a grain of rice
insecticide	something that kills insects
paddy	fields that are flooded with water
polishing	removing the bran layer from rice to make it into white rice
seedling	a very young plant that grows from a seed
spikelet	the flower of the rice plant
starch	gives you energy

terrace a flat area that is cut into a hill

threshing removing the stalks and husks from the rice grains

winnowing blowing away the stalks and husks from the grains

Index

Picture credits: Eye Ubiquitous 28 (David Cumming); Holt Studios 8 (Inga Spence), 9 (S.B.G. McCullagh), 13 (Nigel Cattlin), 14 (Nigel Cattlin), 16 (Inga Spence), 18-9 (D. Donne Bryant), 23 (Inga Spence); Robert Harding 6 (Gavin Hellier), 10 (T. Megarry), 15 (Gavin Hellier); Panos Pictures 7 (Guy Mansfield), 12 (Alain le Garsmeur), 17 (Peter Barker), 22 (Jeremy Hartley). Nick Bailey Photography cover, 3, 5; All other photographs Tim Ridley, Wells Street Studios.

With thanks to Jessica Hopf and Lois Browne.